Christian Schäfer

Assimilation or replacement - a study about Neanderthals and modern humans

GRIN Verlag

Bibliografische Information der Deutschen Nationalbibliothek:

Die Deutsche Bibliothek verzeichnet diese Publikation in der Deutschen National-
bibliografie; detaillierte bibliografische Daten sind im Internet über http://dnb.d-
nb.de/ abrufbar.

Imprint:

Copyright © 2005 GRIN Verlag GmbH
Druck und Bindung: Books on Demand GmbH, Norderstedt Germany
ISBN: 978-3-640-39230-8

This book at GRIN:

http://www.grin.com/en/e-book/41886/assimilation-or-replacement-a-study-about-
neanderthals-and-modern-humans

<u>Essay:</u>

Assimilation or replacement - a study about Neanderthals and modern humans

by Christian Schäfer

Evolutionary Ecology (C-Course)
19.01.2005

Contents

1. Introduction

The Neanderthals lived in Europe and the Near East for at least 250,000 years and they outdared several climate changes. They were capable of surviving in a harsh, cold environment and were well adapted to it – cultural and morphological. Thus, the Neanderthals have been proven to be a successful human kind. But why then did they disappear so quickly and without a trace just between 40,000 and 28,000 yr BP (= years before present) [8]?
One possible answer is that modern humans starting to invade the Near East and Europe out of Africa 45,000 to 40,000 yr BP have outcompeted them, due to higher cultural and mental abilities, using the resources in a more efficient way than the Neanderthals. But is this really true? Have modern humans really had higher abilities? Did they admix with the local Neanderthal populations, integrating the native genes in their gene pool? Or did modern humans not interbreed with them?
And – the big question: were Neanderthals and anatomically modern humans distinct species or just local variants of the same species?
To bring more light into this scenario, these questions will be answered in the following chapters using genetic, morphological and simulation-data that has been brought up by several researchers over the last years.
Answering these fundamental questions also lies in the range of basic needs of human mind: we all want to know where we come from, who was our ancestor and who was it not. To realize which strange ways evolution sometimes takes and to determine what really happened is for sure an exciting thing, and that is exactly what researchers do when they trace human evolution back to the point when Neanderthals and modern humans met in Europe during the last ice age. Only one of them should survive this meeting, and to determine why and how this happened I will start with the description and comparison of the two main characters.

2. Description and comparison of the two main characters

a) Morphology:

A typical Neanderthal man was of short figure - around 160 or 165 cm. There was quite a strong sexual dimorphism as Neanderthal women were around 10 % smaller than men.
In particular the legs were very short compared to the upper part of the body which also showed an enlarged rib cage. This is due to large lungs that made a Neanderthal breathe efficiently. Neanderthals had heavy bones that were very thick and thus quite resistant to mechanical disturbance. They had short thick fingers and arms. Also, the cranium was robust showing thick prominent brow ridges and a flat shape, in particular a low forehead. The face was projecting with a big nose and chinless. The back of the skull showed an occipital bun. Over all, the cranial capacity was very large – the average is around 1500 ccm, but an individual with even around 1800 ccm has been found. From the bone morphology can be deducted that Neanderthals were very muscular and thus weighed a lot compared to their size (around 75 kg at 165 cm). Thus, the typical morphology of a Neanderthal shows a strong adaptation to the rough cold climate in Europe during the ice ages and can be characterized as an extreme sprinter [3], [5].
A typical anatomically modern human is and was taller – with an average of 180 cm today – and shows less sexual dimorphisms as modern human women are around 7% smaller than men. He has quite long legs and a small rib cage. He possesses light, more gracile bones and longer, thinner fingers and arms. The cranium is more gracile in general and has strongly reduced brow ridges. The skull shows a high forehead and is more round in general showing no occipital bun. The face is not projecting with a smaller nose and shorter jaws. It shows a chin. The cranial

capacity is large, but not as large as the Neanderthal one: the average is around 1400 ccm. The body is not very muscular and – at least for primitive people – the weight not that great compared to the size. Anatomically modern humans are thus not well-adapted to cold climates but rather to temperate or subtropical climates and can be characterized as generalists or long-distance runners [3], [5].

b) Cultural abilities:

The Neanderthals were no primitive, culture-less people. In fact, they probably had a language indicated by a fossil hyoid showing strong similarity to ours. Especially the hole for the nerves controlling the tongue is as big as ours suggesting that the tongue's flexibility was very high. Anatomic studies show that the larynx was built the same way as ours. All these features would make no sense unless it gave the Neanderthals the advantage that language and effective communication offers. This advantage far outweighs any disadvantages of spending resources on developing such an anatomy and the added danger of choking. Also, Neanderthals had a highly developed weapon culture, ranging from hunting spears with stone tips to scrapers and hand-axes. Neanderthals were able to use fire and wore clothes that they made from the skin and fur of their prey. They lived in caves and sometimes built small temporary housings. They had a strong social company which is indicated by fossils that show heavy but healed bone-injuries and old people without teeth that were fed by younger group members. Neanderthals were also used to spiritual thinking as they buried their dead in special places with elaborate burial objects like flowers and decorated tools. Thus, they were a folk with both culture and tradition [12].

Anatomically modern humans had all the abilities mentioned above but they had some additional cultural achievements. These were in particular the development of abstract art (cave paintings, music instruments, pearl chains out of shells and the abundant use of red ochre) and better hunting weapon and tool technologies (complex bone technology and multiple-component missile heads). Good examples are the approximately 30,000 year old ivory flute from Germany, the perforated snail shells from Blombos Cave in South Africa which are about 77,000 year old or the cave paintings from Chauvet, France, that are about 30,000 years old [8], [13]. All these things show that perhaps they had a spiritual advantage over the native Neanderthals. Thus, anatomically modern humans show an advanced status of cultural level compared to Neanderthals.

This difference perhaps stems from the better conditions that anatomically modern humans found in the subtropical Africa whereas Neanderthals, although perhaps capable of these things, just didn't have the adequate environment to realize them. Another possible kind of view is that anatomically modern humans had a more flexible mind giving them these cultural advantages, maybe because they were not extremely adapted to a harsh environment and thus not "programmed" to just survive like the Neanderthals. Today we cannot distinguish whether the first or the latter conclusion is right because we unfortunately cannot test a Neanderthal's ability to perform art. Additionally, some artefacts that were found and are associated with art cannot clearly be assigned to modern humans and might be due to Neanderthal art [1]. Also, there is evidence that Neanderthals started manufacturing art when they started to get in contact with anatomically modern humans [8]. Whether they just copied it without understanding it or whether they got the spark and started doing it themselves is not distinguishable today.

c) Dispearsal areas:

The Neanderthals occupied primarily South-Western, Middle and South-Eastern Europe, but fossil finds indicate that they were also present in the Near East [8].

The first fossils of anatomically modern humans were found in South and East Africa, but they soon spread over North Africa to the Near East invading Europe and Asia. Later on, Australia and the American continent were colonized by them. During this population movement process all other archaic human populations like the Neanderthals or *Homo erectus* disappeared all over the world [8].

d) Time span of existence:

The first skeletons with some Neanderthal features are about 300,000 years old. Later fossils, especially from 150,000 yr BP on, show all characteristic features of a typical Neanderthal and thus mark the beginning of existence of this human form in Europe. The last typical Neanderthal skeletons are 28,000 years old and have been found south of the Ebro valley in Southern Spain. This fossil evidence shows that this group must have died out around that time. Thus, Neanderthals and Neanderthal-like humans have existed for at least 250,000 years which indicates that they were a successful human form [12].

The oldest anatomic modern human fossils were found in Ethiopia and are dated back to 160,000 yr BP [8]. This old *Homo sapiens* sub-group is called *Homo sapiens idaltu* [13]. Anatomically modern humans exist until today.

3. Who was the last common ancestor of Neanderthals and modern humans and where did he live?

a) Genetic evidence:

Several mitochondrial DNA (mtDNA) analysis of Neanderthal fossils, Cro-Magnon fossils and current modern humans have shown that the last common ancestor of both Neanderthals and modern humans must have lived between 365,000 and 853,000 yr BP with a best estimate of around 610,000 yr BP [10].

b) Fossil record:

My favourite theory is that the last common ancestor of Neanderthals and *Homo sapiens* lived close to the time of *Homo antecessor* and was closely related to this species. A fossil of this species has been found in the north of Spain and its age is estimated of around 780,000 to 980,000 years. This specimen shows features of both anatomically modern humans and Neanderthals and lived close to the virtual common ancestor that was suggested by the mtDNA analysis mentioned above. Thus, this species might be the ancestor of the last common ancestor or even be *the* last common ancestor of the Neanderthals and modern humans. Another skeleton that has been classified as this species has been found in North Africa and thus supports this theory because *H. antecessor* seems to have existed both in Europe and in Africa – the two potential emergence locations of the Neanderthals and modern humans [4].

Other theories suggest that *Homo heidelbergensis* which was first found in Mauer, Germany, was the last common ancestor. *H. heidelbergensis* lived around 600,000 yr BP [14]. However, at least for the European version of this species this theory seems to fail because the German *H. heidelbergensis* shares lots of traits with the Neanderthals and *H. antecessor*, and is thus likely to be directly descended from *H. antecessor* heading to become a Neanderthal. Also, other specimens of this species are found in Europe and this fact supports the theory that the European *H. heidelbergensis* is an early ancestor of the direct ancestors of the Neanderthals that are sometimes called pre-Neanderthals. On the other hand, the African *H. heidelbergensis* seems to be slightly different from his European kinship. He has less Neanderthal-like features and seems to be a missing link to the later arising species *H. sapiens* [12].

The African *H. heidelbergensis* and several other fossil finds which are located in Africa and fall roughly into the period of 600,000 to 200,000 yr BP are widely summarized in the species-collage "archaic Homo sapiens" and sometimes called "late *Homo ergaster/erectus*". It is likely that anatomically modern humans are descended from one of these "archaic Homo sapiens"-

populations who lived in Africa and that their ancestress was probably *H. antecessor*. The development from these archaic forms to *H. sapiens* was slow but steady (later fossils of this collage show more modern traits than older ones) indicating good living conditions in Africa and therefore a good environment to develop big brains and all other complex features that current humans call their own [12].

4. Description of the main issue

a) Presentation of the four hypothesis concerning the transition from *H. ergaster/erectus* to *Homo sapiens*:

The first hypothesis I will describe is called "Out of Africa"-model. In this scenario anatomically modern humans evolved in Africa from some late *H. ergaster* populations and migrated to Europe and Asia. When they arrived, according to this model, the local human forms were replaced without hybridization and thus no genes from these earlier human forms exist in current humans.

The second scenario is called "Hybridization and assimilation"-model. This hypothesis predicts also that modern humans evolved in Africa and migrated to Europe and Asia. But they interbred with the local populations though the newcomers replaced the natives largely. So, if we follow this scenario, some genes from the archaic populations have entered the modern human gene pool and could possibly have persisted there until today.

The third hypothesis says that modern humans evolved concurrently in Europe, Asia and Africa and is called "Multiregional evolution"-model. A requirement in this scenario is that sufficient gene flow existed between the populations so that they could remain one species. Thus, according to this theory, all local modern human populations have mixed up gene pools consisting out of all possible genes that came up since the development of *H. sapiens* out of *H. erectus/ergaster*. But since the populations in those times were very small and they lived very far away from each other it is nearly impossible that sufficient gene flow could have taken place to make it possible for them to remain one species. So, by just thinking about it without doing any research, this scenario is rather unlikely. And, as I will describe later on, fossil and molecular evidence confirm these doubts.

The last scenario I will describe is called "Candelabra"-model. It is called this way because the phylogeny looks like a candelabra. This hypothesis says that *H. sapiens* evolved independently from local *H. erectus/ergaster* populations. Thus, no gene flow between the populations took place and each current local modern subpopulation is derived from one archaic ancestor subpopulation. Consequently, all regional current human populations should carry genes that are quite different from each other as they are descended from isolated archaic subpopulations. Again, when you think about the basic assumptions of this scenario, you come to the point where you have to conclude that it is highly unlikely that isolated human populations should evolve nearly the same way over more than one million years and that they – in spite of this fact – remain one species. Thus, this scenario can be seen as completely unlikely. And again, molecular and fossil evidence speak their part and thereby confirming these doubts later on [4]. Thus, only the first two hypothesises remain for a reasonable discussion about whether Neanderthals were replaced or assimilated by anatomically modern humans.

b) Integration of "assimilation" and "replacement" into these models:

"Assimilation" would mean that the genes of some Neanderthals have entered the gene pool of our ancestors – anatomically modern populations that evolved in and came from Africa. The theory says that when modern humans invaded Europe and met the native populations, gene flow between the two human forms took place as they interbred. But since this interbreeding did not take place often and the newcomers had an at least slight fitness advantage over the Neanderthals, current Europeans have no visible phenotypic similarity to Neanderthals and only a few Neanderthal genes in their genome. In this kind of view, Neanderthals and anatomically modern humans were local variants of the same species. As we can see these predictions are congruent with the "Hybridization and assimilation"-model described above. "Replacement" means that no interbreeding took place, and if it did, no viable or infertile offspring resulted from it. This theory says that anatomically modern humans from Africa started invading Europe 45,000 to 40,000 yr BP and ruled the Neanderthals out during this process. Thus, no genes of the native populations are present in current Europeans. In this scenario, Neanderthals and anatomically modern humans are different species. These predictions are the same as they are in the "Out of Africa"-model that I have characterized above.

As a conclusion I can say that by doing research on these issues, it can be decided if the two human forms interbred, if the "Hybridization and assimilation"- or the "Out of Africa"-model is true and if Neanderthals and anatomically modern humans were two distinct species or not.

5. Molecular/genetic research

a) mtDNA comparison of Neanderthal fossils, modern human remains and current modern humans:

The first successful Neanderthal mtDNA extraction was performed on the first ever found Neanderthal: the specimen from Feldhofer Cave in the Neander valley, Germany that was found in 1856. The researchers sequenced the hypervariable regions I and II (HVR I and HVR II) after they had extracted the mtDNA from the fossil bones. When they compared these regions of the Neanderthal mitochondrial genome to those ones of modern human remains, they found great differences between them. A first conclusion was drawn that Neanderthals did possibly not contribute to the modern gene pool and that the time of separation of the two human forms was about 317,000 to 741,000 yr BP [6]. But as only one Neanderthal specimen was examined there was little power to discriminate between the alternate hypotheses and data from more Neanderthal specimens would have to be added.

For example, two labs examined the mtDNA of the remains of a Neanderthal child that were found in the Caucasus – the region where modern humans probably entered Europe. The age of the Neanderthal was dated on around 30,000 yr BP by radiocarbon methods. After checking the preservation status of macromolecules in the fossil by determining the mass of collagen-type debris in the used rib-fragment which gave a satisfying result the researchers checked the elemental composition of the collagen. It showed satisfying ratios of carbon and nitrogen indicating a low level of diagenetic modification. After amplifying two overlapping fragments of the HVR I region by PCR (= polymerase chain reaction) the researchers gained enough product to sequence a 345 basepair (= bp) fragment. The result was that the difference from the Neanderthal sequence to current modern humans was much bigger (22 differences) than to the Neanderthal of Feldhofer Cave (12 differences). Also, the two Neanderthals share 19 bp-substitutions relative to the modern human reference sequence. One lab had problems with contamination of modern mtDNA but the samples that stayed pure during the examination gave congruent results with the other lab. The level of pairwise difference between the two Neanderthals is comparable to that of a random sample of 300 current Africans, but less than that seen in current populations of the great apes. The conclusion from these results is that there was very low or no gene flow between modern humans and Neanderthals as the mtDNA sequences of the two Neanderthals who lived 2,500 km away from each other are very similar and the genetic difference between the two Neanderthals and all modern human populations is equal (the average pairwise differences between the Neanderthals and 300 Africans, Mongoloids and Caucasoids showed this) [10].

Later, research on more Neanderthal and anatomic modern human mtDNA was done on fossils that were widely spread over Europe. The use of modern human fossils should make sure that Neanderthal contribution to our gene pool has not been lost by genetic drift. These results reinforced the ones that I described above: all Neanderthal sequences showed no similarity to the ones of modern humans but showed strong similarity to previously published Neanderthal sequences [11].

But still, in this scenario only a Neanderthal contribution to our current gene pool exceeding 25% can be statistically excluded which is a very high value. Also, in this research only the maternal lineage is examined (mtDNA), not the paternal one (Y-chromosomal DNA) [11]. Although it seems unlikely, there could have been differences in the mate choice between males and females: maybe the robust-looking Neanderthal women didn't attract modern human males so much, but the other way round (Neanderthal men and modern human women) were fitting much better, as the muscular and robust Neanderthal men maybe seemed attractive for the rangy modern human women, which in reverse were very attractive for Neanderthal men. So there could have been a lot "hidden" interbreeding, as this described scenario could only be detected by Y-chromosome analysis because mitochondria are only given by mothers to their offspring.

Thus, summarizing the molecular studies, these results show a strong bias to the theory that Neanderthals and modern humans did not interbreed during the range expansion in Europe of the latter. This indicates that these two human forms were probably distinct species. But still, these results are not satisfying enough to confirm this theory as there are some doubts and values that are not clear enough. Maybe more research on preserved DNA from other skeletal remains will give rise to a clearer result.

6. Archaeological and palaeontological research

a) Comparison of the morphology of skeletal remains of different ages and parts of the world:

Today we classify the Neanderthals on the basis of the further above mentioned morphological characteristics. Fossil finds of this typical human form are found everywhere in Europe except in Northern and North-Eastern Europe. The rather pre-Neanderthal fossils are around 300,000 to 150,000 years old and are found in Steinheim, Germany, Swanscombe, England and the Pyrenees, France, e.g. They show many typical Neanderthal features but don't include all of them so that they are undoubtedly the direct ancestors of the later typical Neanderthals. An exceptionally high Neanderthal-fossil-density is found in France and most of the skeletal remains there are dated back to 150,000 to 30,000 yr BP. Some examples of Neanderthal fossil finds in Europe are Spy and La Chapelle in France, Ponthewydd in Great Britain, Monte Circeo in Italy, Gibraltar in Spain, Krapina in Serbia, and of course, the Neander Valley in Germany. But Neanderthal remains are also found in the Near East and Asia like in Shanidar in Iraq, Amud in Syria, Tabun & Kebara in Israel and even in Teshik-Tash in Kazakhstan. In Asia the oldest typical Neanderthal fossils have turned out to be up to 130,000 years old and in the Near East around 80,000 years. They all show a very similar morphology and the specimens in the East show only slight regional morphological differences compared to their western kinship. But Neanderthal fossils are not found south of Israel (in the East) and Gibraltar (in the West) which therefore seem to be a southern dispersal border for this human form [10], [11], [12]. Anatomically modern humans first appeared in Ethiopia 160,000 yr BP and fossil finds from Singa in the Sudan and Israel are around 120,000 and 100,000 years old, respectively. Even these early skeletons show nearly no differences to current human skeletons. The only difference is that the teeth of the earliest modern humans were bigger indicating a stronger use for chewing harder food than we do today – maybe they ate more uncooked plants [5], [8]. Later fossil evidence – from 100,000 years up till now – comes from South Africa and all other parts of Africa which makes us confident that modern humans first colonized Africa before they did anything else, like invading Europe. The first modern human skeletons at the Eastern border of Europe appear just 45,000 yr BP. Other skeletons can be found 40,000 to 30,000 yr BP which appear to be the younger the more western they are located in Europe. The fossils of these humans are called Cro-Magnon fossils, according to the location of the first fossil evidence for this human form in France. This tells us that modern humans progressively invaded the territory of the Neanderthals and started interacting with them. For the sake of completeness, I want to mention that modern humans started migrating to Asia around 60,000 yr BP, they conquered South-East Asia and Australia 50,000 to 60,000 yr BP and colonized the American continent 35,000 to 15,000 yr BP [2], [8].

But the discrimination between Neanderthal and modern human skeletons is not always that easy. Especially some fossils of the time-span from 100,000 to 80,000 yr BP in the Near East and 40,000 to 24,500 yr BP in Europe have given rise to several discussions. The skulls of some specimens show some robust features like prominent brow ridges and a diminished chin, but overall they show clearly modern human traits like the shape of the skull, the jaws, the teeth and many other characteristics. Today these skeletal remains are classified as modern humans with "Neanderthal affinites". Other fossil remains show some modern human features but overall resemble Neanderthals – these specimens are referred to as "advanced Neanderthals".

However, the most controversial fossil evidence comes from Lapedo Valley, Portugal. A well-preserved skeleton of a four year old boy has been dug out in 1998 and has been buried 24,500 yr BP – this is roughly 3,500 years after the Neanderthals have disappeared. But still the remains of this child show strong evidence for Neanderthal traits – mixed with modern human features! The arm bones, chin, jaw and small front teeth resembled modern humans of those times, but the stocky torso, the short legs and the muscle attachment scars were Neanderthal-like. Some scientists interpret this fossil as a proof for the genetic mixing of Neanderthals and modern humans – in this case the child would have been a hybrid. Others say that this individual was just a chunky early modern human indicating the variation among our ancestors 24,500 yr BP. These debates are called "Neanderthal wars" and persist until today. Eventually, this discussion could only end if it was possible to extract DNA out of the remains of this Portuguese child that has been buried with numerous burial objects and red ochre and to find out if it shows mixed genes between modern humans and Neanderthals or not [7].

Overall, the fossil evidence indicates that modern humans and Neanderthals evolved in different environments – the Neanderthals in Europe and the Near East and modern humans mainly in Africa. They met each other after several hundred thousand years of isolation, but the fossil evidence is not explicit enough to draw clear conclusions from it – it is not distinguishable today if the two human forms interbred or not and if they thus display the same or different species.

<u>b) Comparison of the cultural remains of different ages and parts of the world</u>:
Scientists classify archaic technologies appearing in Europe 300,000 yr BP as "Mousterian" and more modern ones as "Aurignacian" technologies. Roughly all Aurignacian technologies are associated with modern human fossils and the Mousterian ones came up with the first Neanderthal-like fossils and accompany the Neanderthals till the end of their existence. Thus, it seems to be true that the Aurignacian culture evolved within modern humans and that Neanderthals were the creators of Mousterian technologies [9].

The Aurignacian culture can be divided into two subclasses – the "classic" Aurignaic and the "archaic" or "proto" Aurignaic. Both stem from Ahmarian and Emiran technologies which can be found in the Near East and are 47,000 to 45,000 years old.

The classic Aurignaic is represented at Aurignaic itself and is marked by a lot distinctive tool forms like typical nosed and carinated scraper forms, heavily edge-trimmed Aurignacian blades and split base bone and antler spear-head forms. Cultural remains belonging to these technologies can be found in western, central and south-eastern Europe and even in the adjacent areas of the Near East. The technologies astonishingly show nearly no variation all over Europe. Its origin is located in the eastern Mediterranean region and in south-eastern Europe around 40,000 yr BP. The archaeological record indicates that the classic Aurignaic appeared in other European areas 38,000 to 34,000 yr BP and shows an abrupt break with the preceding Neanderthal technologies. The route people manufacturing tools and art with these technologies took is called the "northern" route because it led them north of the river Danube through Europe.

The archaic or proto Aurignaic represents a different pattern of technology from that of the classic technologies. It is marked by some distinctive tool forms like small, carefully shaped bladelets which probably served as tips and barbs of composite spear- or arrowheads. These technologies can be found along the Mediterranean coast of Europe, in north-eastern Italy and at the Atlantic coast of Spain. Again, they show a sharp break with the older Neanderthal technologies. The origin of this industry is located in the Near East 38,000 to 40,000 yr BP. The dispersal of tools from these technologies shows a pattern that is called the "southern" route because it mainly involves southern Europe.

An interesting point is that some Neanderthal subpopulations started manufacturing Aurignaic-like tools and art after modern humans had invaded their areas. An example comes from the Pyrenees in Spain where 38,000 year old Neanderthal fossils are associated with penetrated shells, red ochre and other newly gained cultural features that are slightly different but still astonishingly similar to Aurignaic technologies. At this time modern humans already lived in this area, so a cultural knowledge transfer cannot be ruled out. Other interpretations of these facts say that maybe a yet unknown reason is responsible for a sudden cultural development among Neanderthals that was not influenced by modern humans [1], [8].

Summarizing these insights, I can say that it is not distinguishable today, if Neanderthals had not the cognitive capabilities to develop cultural abilities like modern humans did with the Aurignaic and if they just copied these things from the newcomers without understanding it or if they were potentially able to do it and just couldn't realize it because of environmental constraints. The cultural remains unfortunately can tell us nothing about whether Neanderthals and modern humans belonged to the same species or not. But what is sure is that modern humans brought with them advanced technologies that they developed in the Near East and which made it easier for them to invade Europe and eventually the whole world. This was maybe one or *the* reason why Neanderthals were finally driven to extinction. This development might be due to luck or to extraordinary mental abilities that no human form before has gained and thus made it possible for modern humans to become the one and only human form on this planet.

7. Simulation studies

a) What are they based on?

Older models could only statistically exclude a contribution of Neanderthal mtDNA to the current human gene pool exceeding 25%, in spite of the great differences of the sequenced DNA between Neanderthals and both current humans and Cro-Magnon people. The old population development models were quite simple: one said that both the Neanderthal and the modern human population was unsubdivided and mixed instantaneous. The other model included the assumptions of the first one but added that the modern human population started growing exponentially before the admixture event occurred and continued doing so after it. The newly developed, more complex model is called "progressive range expansion of modern humans into Europe" and includes spatial information where the modern human population occupies a different range than the Neanderthal population before the admixture started. Here, admixture is progressive and happens because modern humans move into the territory of Neanderthals that becomes smaller and smaller during this process until Europe is completely settled by modern humans. Further differences to the previous approaches include the subdivision of Europe into small territories potentially harbouring two subpopulations – a modern human and a Neanderthal deme; the local population size is logistically regulated for both human forms; there is competition between Neanderthals and modern humans which leads to a victory of the latter ones due to their higher carrying capacity; thus, the admixture is also progressive and occurs only in the narrow strip at the front of the expanding modern human subpopulations; integration of coalescent simulations which are used to estimate the likelihood of different rates of admixture, given that Neanderthal mtDNA is not observed in current Europeans.

Thus, the range expansion and the admixture process will depend on several parameters: the carrying capacity of the local subpopulations, their intrinsic growth rate, the amount of gene flow between close demes, the local rate of admixture, the geographic origin of the range expansion and many more. Archaeological and paleodemographic data is used to calibrate the values of these parameters [2].

b) What are the results?

The simulations even under very different starting values (different scenarios) show that under these more realistic conditions a maximum admixture rate of on the average 0.0625% (range 0.02 to 0.09%) is possible when conditioned on the fact that no Neanderthal mtDNA remains in the current European population. Expressed in total numbers, this means that the maximum number of interbreeding events is expected to have been between 34 and 120 during the whole cohabitation period of 12,000 years. These are indeed extremely low values. These figures represent a big improvement compared to the old models that could exclude just 25% or more Neanderthal contribution to the modern human gene pool.

The simulation's results are quite robust to variations in local modern human subpopulation growth rates, levels of gene flow between adjacent modern human demes, the kind of interbreeding (symmetric or only Neanderthal men and modern human women, e.g.), the geographic origin of the modern human range expansion and the size and spread of the population at the source of the range expansion. Thus, if the model is correct, all current European genes can be followed back to a small number of individuals – a bottleneck that occurred 40,000 yr BP. This kind of view is supported by research done on European mtDNA giving the same results.

When local logistic and non-logistic subpopulation growth are compared the results indicate that local logistic growth reduces the probability of losing any Neanderthal genes that may have entered the modern human gene pool. Here, only around 50 fertile interbreeding events can have occurred during the whole cohabitation time. This scenario is very probable as not the whole population invades a new territory. On the contrary, a scenario without logistic growth allows much greater interbreeding rated to give the same results – here, a maximum of 1,850 fertile breeding events (= 1.2% initial Neanderthal input) between the two human forms is possible despite the fact that no Neanderthal mtDNA is present in current Europeans. But this scenario is not a very likely one as it assumes an unlikely population development.

Also, the fact that no Neanderthal mtDNA is present in current Europeans is probably not caused by selection against these genes as they would have been even better adapted to the cold

climate in Europe than the genes of modern humans who evolved in the warmer environment of the Middle East and Africa. Therefore, selection is not a likely cause of the picture we see today. Thus, it is more likely that higher cognitive abilities and/or better technologies drove this process of population change than a selective advantage of modern human mtDNA.
But to make sure that this basic assumption is right, studies on Y-chromosome and nuclear DNA have to be done in the future [2].
This chapter gives us a very clear output of the data that is available so far as on the basis of realistic assumptions and varying starting values a consistent picture is drawn which shows that the amount of interbreeding events between modern humans and Neanderthals must have been very low – in the most realistic scenario around 50 in 12,000 years, but not more than 120. Therefore, it is appropriate to assume these two human forms were actually two different biological species – *Homo sapiens* and *Homo neanderthalensis* – as their offspring was probably sterile or had a drastically reduced fitness. This again suggests that the Neanderthals were completely replaced by modern humans invading their territory, and not assimilated. Thus, Neanderthals don't belong to our ancestors, and as no mtDNA of other archaic Homo-populations from Asia (like late *Homo erectus*) can be found in the current modern human gene pool, an admixture of them with our ancestors can be excluded, too. This favours explicitly the "Out of Africa"-hypothesis and suggests a closely related current human population. The hypothesis is supported by genetic research which indicates that all modern human subpopulations vary astonishingly little in their DNA sequences among each other compared to other primates.

8. Conclusions

As molecular research and the fossil record suggest, the last common ancestor of Neanderthals and modern humans has lived around 600,000 yr BP and was probably located in North Africa and North Spain. This divergence time is long enough to accumulate biological and behavioural differences that finally lead to a speciation process. Also, the different environments the two separated populations lived in gave rise to different selection pressures which surely formed their genetic and cultural development. The harsh climate in Europe during the ice ages forced the Neanderthals to develop morphological and behavioural specialities that maybe did not let them be innovative enough to invent the advanced techniques that modern humans in the warm and moderate climate south of Europe could develop. This probably was the reason why the Neanderthals finally vanished into thin air. The molecular, the archaeological and the palaeontological research give similar results which all lead to a bias that the Neanderthals were replaced and not assimilated by modern humans invading their territory, but some arguments are not that sure and sometimes there is evidence for the contrary. But the newest study – a simulation study which integrates all former results – indicates that an admixing between these two human forms is extremely unlikely and that thus Neanderthals and modern humans were distinct species. Therefore, this study enforces former assumptions and leads to an establishment of the "Out of Africa"-hypothesis.
As a conclusion, I can say that modern humans invading Europe 40,000 to 28,000 yr BP probably replaced the Neanderthals completely and are thus the only ancestors of current Europeans.

9. References

[1] Bahn, Paul G.: Neanderthals emancipated. Nature, Vol. 394, p. 719-721 (1998).

[2] Currat, Mathias & Excoffier, Laurent: Modern Humans Did Not Admix with Neanderthals during Their Range Expansion into Europe. PLoS Biology, Vol. 2, Issue 12, p. 0001-0011 (2004).

[3] Fleagle,, J.G.: Primate Adaptation and Evolution. London: Academic Press (1999).

[4] Freeman, S. & Herron, J.C.: Evolutionary Analysis, Third Edition (2004).

[5] Jurmain, R. & Nelson, H.: Introduction to physical anthropology. St. Paul, Minneapolis: West Publishing Company (1994).

[6] Krings, Mathias et al.: Neandertal DNA Sequences and the Origin of Modern Humans. Cell, Vol. 90, 19-30 (1997).

[7] Lubenow, Marvin L.: Lagar Velho 1 child skeleton: a Neandertal/modern human hybrid. CEN Technical Journal 14(2) (2000).

[8] Mellars, Paul: Neanderthals and the modern human colonization of Europe. Nature, Vol. 432, p. 461-465 (2004).

[9] Mellars, Paul: The fate of the Neanderthals. Nature, Vol. 395, p. 539-540 (1998).

[10] Ovchinnikov, Igor V. et al.: Molecular analysis of Neanderthal DNA from the northern Caucasus. Nature, Vol. 404, p. 490-493 (2000).

[11] Serre, David et al.: No Evidence of Neandertal mtDNA Contribution to Early Modern Humans. PLoS Biology, Vol. 2, Issue 3, p. 0313-0317 (2004).

[12] Spektrum der Wissenschaft, Dossier: Die Evolution des Menschen (2000).

[13] www.wissenschaft-online.de

[14] www.homoheidelbergensis.de